By

An Easy Steps Math series book

Books in the Easy Steps Math series

Fractions
Decimals
Percentages
Ratios
Negative Numbers
Algebra
Master Collection 1 – Fractions, Decimals and Percentages
Master Collection 2 – Fractions, Decimals and Ratios
Master Collection 3 – Fractions, Percentages and Ratios
Master Collection 4 – Decimals, Percentages and Ratios

More to Follow

Contents

Introduction

This series of books has been written for the purpose of simplifying mathematical concepts that many students (and parents) find difficult. The explanations in many textbooks and on the Internet are often confusing and bogged down with terminology. This book has been written in a step-by-step 'verbal' style, meaning, the instructions are what would be said to students in class to explain the concepts in an easy to understand way.

Students are taught how to do their work in class, but when they get home, many do not necessarily recall how to answer the questions they learned about earlier that day. All they see are numbers in their books with no easy-to-follow explanation of what to do. This is a very common problem, especially when new concepts are being taught.

For over twenty years I have been writing math notes on the board for students to copy into a note book (separate from their work book), so when they go home they will still know how the questions are supposed to be answered. The excuse of not understanding or forgetting how to do the work is becoming a thing of the past. Many students have commented that when they read over these notes, either for completing homework or studying for a test or exam, they hear my voice going through the explanations again.

Once students start seeing success, they start to enjoy math rather than dread it. Students have found much success in using the notes from class to aid them in their study. In fact students from other classes have been seen using photocopies of the notes given in my classes. In one instance a parent found my math notes so easy to follow that he copied them to use in teaching his students in his school.

You will find this step-by-step method of learning easier to follow than traditional styles of explanation. With questions included throughout, you will gain practice along with a newfound understanding of how to complete your calculations. Answers are included at the end.

Chapter 1

Decimal Basics

A decimal is also known as a **decimal fraction**. This is because it deals with parts of numbers that are **less than one,** just like fractions. However instead of writing numbers over each other, they are written beside a decimal point (.) that looks just like a period or full stop.

The numbers to the left of the decimal point are whole numbers, also called **integers**. The numbers to the right of the decimal point are just referred to as **decimal values** or **decimal numbers**.

E.g.

Decimal Point

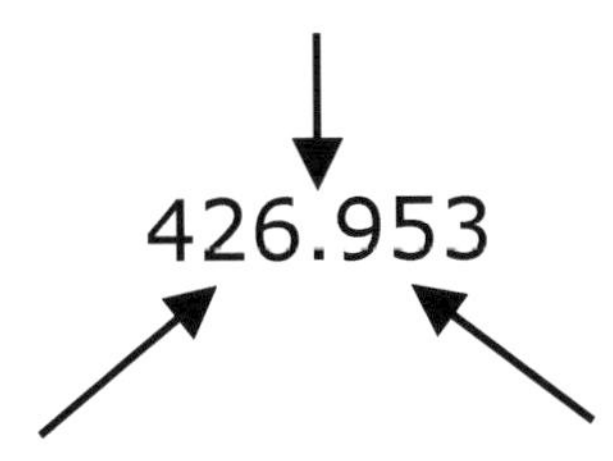

Integers **Decimals Numbers**

The whole number and the decimal numbers are separated by the decimal point.

You saw in the Easy Steps Math Fractions book, that a whole number could be written as a fraction by putting it over 1, hence 3 becomes $\frac{3}{1}$. Well something similar can be done for decimals. All whole numbers can be written as decimals by putting a decimal point and then a zero or two or three after the last number.
Therefore,

3 can be rewritten as 3.0, 3.00, and 3.000…
25 is the same as 25.0, 25.00 …
4398 is the same as 4398.0, etc.

The importance of this becomes obvious a bit later when you do calculations with decimals.

Chapter 2

Place Values

Each digit in a number, whole or decimal, has a specific place value. It's position; relative to the decimal point is what determines its place value. For instance, looking at the number 426.953 you know the first three digits are four, two and six, which makes four hundred and twenty-six and the last three digits are nine, five, and three.

(Note that the digits to the right of a decimal point are expressed individually and not as a larger number).

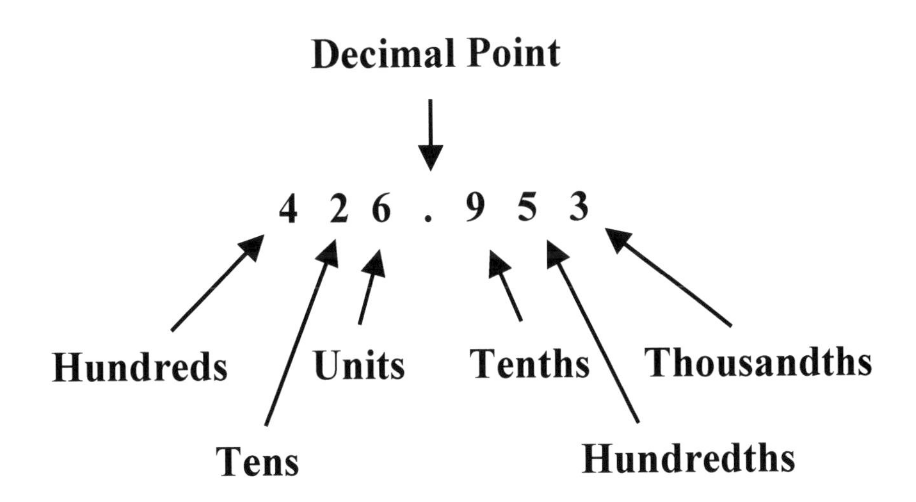

These place values continue on to infinity in both directions

Each digit in the above number has it's own value based on how far it is from the decimal point. The numbers to the left of the decimal point are easy because you actually say the value when you say the number. Four hundred, (which is obvious) and twenty, (which is two tens) and six (which is six units).

To the right of the decimal point, the value of:

- the first digit is nine tenths. As a decimal it is written as 0.9, as a fraction it is written as $\frac{9}{10}$.
- the second digit is five hundredths. As a decimal it is written as 0.05, as a fraction it is written as $\frac{5}{100}$.
- the third digit is three thousandths. As a decimal it is written as 0.003, as a fraction it is written as $\frac{3}{1000}$.

If you were to convert the number 426.953 to a fraction it would be $426\frac{953}{1000}$.

Remember that when you say the decimal, you would say four hundred and twenty-six point nine five three.

Try the following questions.

What is the value of the indicated digit in the following? Write your answer as a fraction.

a) 326.153 The value of 5 is

b) 459.364 The value of 3 is

c) 25.63 The value of 2 is

d) 123.456 The value of 6 is

e) 2.3578 The value of 8 is

Now write these fractions as decimals

f) $\frac{3}{10} =$

g) $\frac{7}{100} =$

h) $\frac{15}{100} =$

i) $2\frac{23}{100} =$

j) $15\frac{145}{1000} =$

You may have noticed the pattern that exists. When you convert decimals to fractions, the number of zeros in the denominator of the fraction is the same as the number of positions to the right of the decimal point of the decimal.

I.e.

0.15 has two decimal places therefore it will have two zeros in the denominator so it becomes $\frac{15}{100}$.

0.026 has three decimal places therefore it will have three zeros in the denominator so it becomes $\frac{26}{1000}$.

0.0037 has four decimal places therefore it will have four zeros in the denominator so it becomes $\frac{37}{10000}$. And it continues.

Comparing decimals is important to learn because at times you need to determine which decimal values are larger and which ones are smaller. Like money for instance.

Which is larger \$5.25 or \$5.52? You know the second one, \$5.52, is larger.

Money, (dollars and cents) is called decimal currency, so thinking of decimals as money can sometimes help with which values are larger and which values are smaller. Thinking of decimals as money can help with the calculations of decimals generally.

If you take away the dollar signs (\$), can you still work out which is the larger decimal, 5.25 or 5.52? What about 3.70 or 3.07?

Put the following in ascending (lowest to highest) order.

a) 4.56, 6.45, 5.64, 4.65, 5.46, 6.54

b) 3.7, 3.40, 3.75, 3.41, 3.73, 3.5

If some of the decimal numbers have less decimal places shown, zeros can be added to make the number of decimal places the same, hence 4.9 is the same as 4.90.

E.g. If you need to put 7.698, 7.7, 7.2467, 7.35, in ascending order, you can add the necessary number of zeros to make the number of place values the same for all the decimal numbers. In this case it is four decimal places.

Therefore 7.698 becomes 7.6980, 7.7 becomes 7.7000, 7.2467 stays as it is and 7.35 becomes 7.3500.

To order these you can see that 7.2467 would come first, then 7.3500, then 7.6980, and lastly 7.7000.

The final answer is therefore, 7.2467, 7.35, 7.698, 7.7
Put the following in descending (highest to lowest) order

c) 4.677, 4.706, 4.71, 4.67, 4.6

d) 3.616, 3.116, 3.661, 3.66

e) 2.45, 2.37, 4.983, 3.102, 1.874, 1.876, 3.098, 4.5.

Note that when you're putting decimals in ascending or descending order, don't forget to take the whole numbers into account also.

The **greater than** symbol **(>)** or **less than** symbol **(<)** can also be used to state which decimal number is larger or smaller. These symbols are usually read from left to right so the greater than symbol has a wider opening on the left than on the right, and the less than symbol has a smaller opening on the left (actually it has no opening). Therefore using these symbols, 5.52 > 5.25 (5.52 is greater than 5.25) and 3.07 < 3.70 (3.07 is less than 3.70).

f) Which of the following is incorrect?

A 6.646 > 6.466
B 6.646 > 6.664
C 6.646 > 6.644
D 6.646 > 6.464
E 6.646 > 6.64

g) Which of the following is correct?

A 0.7149 > 0.7419
B 0.7914 < 0.7491
C 0.7194 > 0.7149
D 0.7941 < 0.7914
E 0.7194 > 0.7419

Chapter 3

Rounding Decimals

Many decimal numbers have more decimal places than what is needed for a calculation. Some of these are **recurring decimals** and others are **non-recurring**. In either case, there are too many digits to work with and calculations become difficult. To solve this difficulty, decimal values are rounded up or down.

For example, the Australian government eliminated all one and two cent coins in 1991. So now if you buy something and its price is $3.41 or $3.42, then the amount you pay is rounded down to $3.40. If the price is $3.43 or $3.44, then the amount you pay is rounded up to $3.45. Prices that end in 6 cents or 7 cents are rounded down to the nearest 5 cents and prices that end in 8 cents or 9 cents are rounded up to the next 10 cents. So anything with the price of $3.98 or $3.99 will actually cost $4.00.

With decimal numbers something similar is done. If a question asks you to round to a certain number of decimal places, you need to look at the next decimal place to decide what to do.
If that next decimal place is a **0, 1, 2, 3 or 4**, then you **round down** which means you just drop all digits after the one you want to keep.
If it is a 5, 6, 7, 8 or 9, then you round up which means the one you want to keep goes up one digit and all the ones after are dropped.

Examples

a) Round 17.2395 to **2** decimal places

Step 1. As the question asks you to round to 2 decimal places, you need to look at the next decimal place, i.e. the third digit.

17.2395

Step 2. In this case the third digit is a 9 therefore you must round up.

Step 3. Change the 3 up to a 4 and drop all other digits after it. So the answer is 17.24

b) Round 12.13224 to **3** decimal places

Step 1. As the question asks you to round to 3 decimal places, you need to look at the next decimal place, i.e. the forth digit.

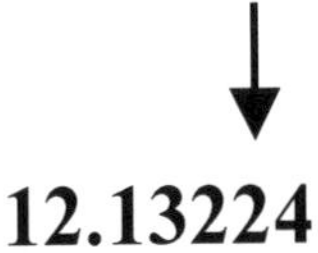

12.13224

Step 2. In this case the forth digit is a 2, therefore you must round down.

Step 3. Just drop all the digits after the one you want to keep. So the answer is 12.132

Remember, if the decimal value of the digit <u>after</u> the one you want to keep is a 0, 1, 2, 3 or 4, you would round down. If the value is a 5, 6, 7, 8 or 9, then you would round up.

Try some of these questions

Round each of the following to 1 decimal place

a) 18.62 b) 7.58 c) 0.55 d) 9.44 e) 7.92 f) 7.98

Round each of the following to 2 decimal places

g) 5.792 h) 12.5682 i) 32.4573 j) 400.2648

Round each of the following to the nearest whole number

k) 26.8 l) 5.29 m) 83.5 n) 121.11 o) 75.75

Round each of the following to the nearest dollar

p) $26.85 q) $15.29 r) $423.50 s) $21.41

Chapter 4

Addition and Subtraction of Decimals

When adding and subtracting decimals, you need to follow these steps:

Step 1. Line up the decimal points over one another

Step 2. Insert any zeros after the last digit so all the numbers have the same number of decimal places.

Step 3. Put the decimal point in your answer directly below the other decimal points (do this first) then add or subtract as you would whole numbers.

Examples

a) Find 6.35 + 19.58

Step 1. Line up the decimal points

$$\begin{array}{r} 6.35 \\ +\ \underline{19.58} \end{array}$$

Step 2. Insert any extra zeros if needed
Step 3. Insert a decimal point in the answer space then add

$$\begin{array}{r} 6.35 \\ +\ \underline{19.58} \\ 25.93 \end{array}$$

b) Find 8.54 + 6.3 +3.487

Step 1. Line up the decimal points

$$\begin{array}{r} \mathbf{8.54} \\ \mathbf{6.3} \\ + \underline{\mathbf{3.487}} \end{array}$$

Step 2. Insert any extra zeros if needed

$$\begin{array}{r} \mathbf{8.540} \\ \mathbf{6.300} \\ + \underline{\mathbf{3.487}} \end{array}$$

Insert extra zeros as needed

Step 3. Insert a decimal point in the answer space then add

$$\begin{array}{r} \mathbf{8.540} \\ \mathbf{6.300} \\ + \underline{\mathbf{3.487}} \\ \mathbf{18.327} \end{array}$$

Insert extra zeros as needed

Note that the extra zeros can only be inserted **after** the last digit.

c) Find 8.53 – 3.12

Step 1. Line up the decimal points

$$\begin{array}{r} 8.53 \\ -\ \underline{3.12} \end{array}$$

Step 2. Insert any extra zeros if needed
Step 3. Insert a decimal point in the answer space then subtract

$$\begin{array}{r} 8.53 \\ -\ \underline{3.12} \\ 5.41 \end{array}$$

d) Find 176.85 – 52.5

Step 1. Line up the decimal points

$$\begin{array}{r} 176.85 \\ \underline{-\ 52.5} \end{array}$$

Step 2. Insert any extra zeros if needed

$$\begin{array}{r} 176.85 \\ \underline{-\ 52.50} \end{array}$$

Step 3. Insert decimal point in answer space then subtract

176.85
- 52.50
124.35

Find each of the following **without** a calculator

a) 12.58 + 7.9

b) 9.4 + 15.6

c) 77.81 + 6.3

d) 23.53 – 15.21

e) 216.8 – 148.564

f) 235.6 + 25.45

g) 25.8 + 68.23

h) 236.5 - 45.568

i) 16.89 - 0.116

j) 44.563 - 23.6

Chapter 5

Multiplication of Decimals

When multiplying decimals with whole numbers, you multiply them as though you are multiplying just whole numbers and you place the decimal point after you do all the calculations.

For example, if you were to multiply 13.7 x 9, you would follow these steps:

Step 1. Set out the numbers like a multiplication of whole numbers

$$\begin{array}{r} \mathbf{13.7} \\ \underline{\mathbf{x\ \ 9}} \end{array}$$

Step 2. Ignore the decimal point at this stage and multiply the numbers

$$\begin{array}{r} \mathbf{13.7} \\ \underline{\mathbf{x\ \ 9}} \\ \mathbf{1223} \end{array}$$

Step 3. Count the number of decimal places in **both** parts of the question (this is to be done from right to left). In this case there is only one decimal place in 13.7

Step 4. Count one decimal place **from the far right** of the answer and put in the decimal point. Therefore 1233 becomes 123.3

```
 13.7
 x  9
-----
122.3
```

So 13.7 x 9 = 123.3

Here is another example.

Calculate 9.358 x 12

Following the steps above:

Step 1. Set out the numbers like a multiplication of whole numbers

```
9.358
x  12
-----
```

Step 2. Ignore the decimal point at this stage and multiply the numbers

```
  9.358
  x  12
 ------
  18716
  93580
 ------
 112296
```

Step 3. There are three decimal places in **both** parts of the questions. This is in 9.358

Step 4. For the answer place the decimal point three places from the right. Therefore the answer is 112.296

$$
\begin{array}{r}
\mathbf{9.358} \\
\underline{\mathbf{x\ \ 12}} \\
\mathbf{18716} \\
\underline{\mathbf{93580}} \\
\mathbf{112.296}
\end{array}
$$

So 9.358 x 12 = 112.296

Multiply these decimals and whole numbers.

a) 3.5 x 4

b) 12.3 x 9

c) 10.2 x 8

d) 12.31 x 15

e) 16.46 x 23

f) 0.6548 x 12 round your answer to 3 decimal places.

g) 9.026354 x 7 round your answer to 4 decimal places.

h) 5.67 x 65 round your answer to 1 decimal place.

i) 56.32 x 23 round your answer to 2 decimal places.

j) 6.98 x 15 round your answer to 2 decimal places.

In the situation where you multiply a decimal number with another decimal number, you follow the exact same steps as above. Nothing changes.

Examples,

a) *Calculate* 12.6 x 3.2

Step 1. Set out the numbers like a multiplication of whole numbers

12.6
x 3.2

Step 2. Ignore the decimal points and multiply the numbers

12.6
x 3.2
4032

Step 3. Count the number of decimal places in **both** parts of the question (this is done from the right to left). In this case there is one decimal place in 12.6 and one in 3.2 so two in total.
Step 4. Count two decimal places **from the far right** in the answer and put in the decimal point. Therefore 4032 becomes 40.32

$$\begin{array}{r} 12.6 \\ \times\ 3.2 \\ \hline 40.32 \end{array}$$

So 12.6 x 3.2 = 40.32

b) *Calculate* 4.361 x 3.2

Step 1. Set out the numbers like a multiplication of whole numbers

$$\begin{array}{r} 4.361 \\ \times\ \ 3.2 \\ \hline \end{array}$$

Step 2. Ignore the decimal points and multiply the numbers

$$\begin{array}{r} 4.361 \\ \times\ \ 3.2 \\ \hline 8622 \\ 130830 \\ \hline 139452 \end{array}$$

Step 3. Count the number of decimal places in **both** parts of the question. In this case there are four decimal places in **both** parts

Step 4. For the answer, place the decimal point four places from the right. Therefore the answer is 13.9452

4.361
x 3.2
8622
130830
13.9452

So 4.361 x 3.2 = 13.9452

c) *Calculate* 0.03 x 0.02

Step 1. Set out the numbers like a multiplication of whole numbers

0.03
x 0.02

Step 2. Ignore the decimal points and multiply the numbers

0.03
x 0.02
006
0000
00000
00006

Step 3. Count the number of decimal places in **both** parts of the question. In this case there are four decimal places in **both** parts

Step 4. For the answer, place the decimal point four places from the right. Therefore the answer is 0.0006

```
     0.03
x    0.02
---------
      006
     0000
    00000
---------
   0.0006
```

So 0.03 x 0.02 = 0.0006

Multiply these decimals with decimals

a) 0.4 x 0.8

b) 0.83 x 0.9
c) 2.7 x 3.6

d) 9.46 x 0.22

e) 0.628 x 2.4

f) 5.007 x 6.4 round your answer to 3 decimal places

g) 1.03 x 0.02 round your answer to 3 decimal places

h) 0.004 x 0.06 round your answer to 4 decimal places

i) 25.34 x 78.9 round your answer to 2 decimal places

j) 14.95 x 0.9 round your answer to 2 decimal places

If you were to multiply a decimal by a multiple of 10 i.e. 10, 100, 1000, etc, then all you do is move the decimal point to the right the same number of places as there are zeros in the multiplier. Moving the decimal point to the right is the same as multiplying by 10, 100, etc. Test it on your calculator.

For example,

1. If you were to multiply 23.659 x 100, you would simply move the decimal point two places to the right, so the answer would be 2365.9

2. If you were to multiply 23.659 x 10, you would simply move the decimal point one place to the right, so the answer would be 236.59

3. If you were to multiply 23.659 x 300, you would follow a two-step process.

First you would move the decimal point as though you were multiplying by 100. So you would have 2365.9.

Second you would multiply this answer by 3, (following the multiplying a decimal by a whole number method).

2365.9
x 3
7097.7

Therefore 23.659 x 300 = 7097.7

Occasionally you will get a question like calculate 5.6 x 1000

In this situation, you write in a few extra zeros after the last digit (6), and then move the decimal point to the right three places.

So …

5.6 x 1000 is the same as 5.6000 x 1000 (the extra zeros after the 6 (the last digit) are allowed in this case as it does not change the question).

Now multiply by 1000 by moving the decimal point to the right three spaces.

Remember, the number of zeros in the multiplier is the same as the number of spaces you move the decimal point to the right.

Now you try these questions.

Calculate:

a) 8.36 x 100 =

b) 0.35 x 100 =

c) 4.63 x 1000 =

d) 2.305 x 300 =

e) 3.27 x 7000 =

f) 2.2562 x 300 =

g) 3.9 x 1200 =

h) 0.32 x 25000 =

i) 1.2 x100 =

j) 7.56 x 10000 =

Chapter 6

Division of Decimals

If you remember from the Easy Steps Math Fractions book, the three parts of a division are:

The dividend – the number being divided
The divisor – the number doing the dividing
The quotient – the answer.

These can be set out with a division symbol like so:

$$Dividend \div Divisor = Quotient$$

or in a division box like so:

$$Divisor\overline{)Dividend}^{\,Quotient}$$

or as a fraction like so:

$$\frac{Dividend}{Divisor} = Quotient$$

When you divide a decimal by a whole number the most important part to remember is **to line up the decimal points of the dividend and the quotient.** So it needs to look like this:

$$3\overline{)15.9}^{\;5.3}$$

Notice how the decimal points in the dividend and quotient line up. The best way to make sure this happens is to place the decimal point in the quotient **before** you start dividing.

So if you were given the question *calculate* $15.9 \div 3$, your first step would be to write:

$$15.9 \div 3 = 3\overline{)15.9}^{\;\;\;.}$$

Take note of the decimal point in the quotient space, in the correct location, before any calculations are made. Now you can start working out the numbers.

Here is another example:

Calculate $8.24 \div 4$

Step one: Rewrite the question using the 'division box'.

$$8.24 \div 4 = 4\overline{)8.24}$$

Step 2: Place the decimal point in the quotient space in line with the decimal point already there.

$$8.24 \div 4 = 4\overline{)8.24}^{\;\;.}$$

Step 3: Work out the division as though they were whole numbers.

4 goes into 8 twice, so place a 2 above the 8

$$4\overline{)8.24}\quad \text{quotient: } 2.$$

4 doesn't go into 2, so place a 0 above the 2

$$4\overline{)8.24}\quad \text{quotient: } 2.0$$

4 goes into 24 six times. So put a 6 next to the 0, above the 4.

$$4\overline{)8.24}\quad \text{quotient: } 2.06$$

So $8.24 \div 4 = 2.06$

Divide these decimals with whole numbers

a) $4.1 \div 7$ and round your answer to 3 decimal places.

b) $26.48 \div 5$ and round your answer to 2 decimal places.

c) $2.822 \div 4$ and round your answer to 3 decimal places.

d) $16.5 \div 8$ and round your answer to 3 decimal places.

e) $0.603 \div 2$ and round your answer to 2 decimal places.

f) $1.79 \div 8$ and round your answer to 3 decimal places.

g) $3.3 \div 2$ and round your answer to 1 decimal place.

h) $5.1 \div 6$ and round your answer to 2 decimal places.

i) $8.25 \div 2$ and round your answer to 3 decimal places.

j) $1.964 \div 9$ and round your answer to 2 decimal places.

When dividing a decimal by another decimal, such as $5.342 \div 0.04$, you need to follow these steps before you begin your division.

Step 1. Change the divisor to a whole number. Moving the decimal point to the right of the last digit does this. In our example, the decimal point moves 2 places to the right, making the divisor 4.0

Step 2. Move the decimal point in the dividend the exact same way. In our example it is 2 places to the right, making the dividend 534.2

Step 3. Divide using the method learned above, dividing a decimal by a whole number.

So, our new question will be $534.2 \div 4$ which is the same as $5.342 \div 0.04$, the original question.

The answers for both will be exactly the same.

If you remember from the Easy Steps Math Fractions book, you learned that what you do to one part of the fraction you must do to the other part of the fraction. Since a fraction is just a division, the same thing applies here.

Here is another example

Calculate $10.6 \div 0.005$

Step 1. Change the divisor to a whole number. Move the decimal point in the divisor to the right of the last digit.

Moving the decimal point three spaces to the right changes 0.005 to the whole number 5.0

Step 2. Move the decimal point in the dividend the exact same way.

In this situation, when you move the decimal point three spaces to the right, you can add extra zeros to the dividend so there is someplace for the decimal point to go. This changes 10.6 to the new number 10600.0

Step 3. Divide as usual.

So our new question will be $10600 \div 5$

$$= 5\overline{)10600} = 5\overline{)10600}^{\;2120}$$

The answers are exactly the same between the old and new questions.

Divide these decimals by another decimal.

a) $24.42 \div 0.2$

b) $5.1 \div 0.6$

c) $9.96 \div 1.2$

d) $6.732 \div 0.0012$

e) $1.74 \div 0.0002$

f) $0.14 \div 0.3$ and round your answer to 3 decimal places

g) $25.7 \div 0.12$ and round your answer to 3 decimal places

h) $0.1776 \div 0.0031$ and round your answer to 3 decimal places

i) $4.632 \div 0.005$ and round your answer to 1 decimal places

j) $0.584 \div 0.6$ and round your answer to 2 decimal places

When dividing a decimal with a multiple of 10 you do the opposite to what you do for multiplication. That is you move the decimal point to the left the same number of places, as there are zeros in the divisor. Moving the decimal point to the left is the same as dividing by 10, 100, etc. Test it on your calculator.

For example,

1. If you were to divide 23.659 by 100, you would simply move the decimal point two places to the left, so the answer would be 0.23659

2. If you were to divide 23.659 by 10, you would simply move the decimal point one place to the left, so the answer would be 2.3659

3. If you were to divide 23.659 by 300, you would follow a two-step process.

First you would move the decimal point as though you were dividing by 100. So you would have 0.23659.

Second you would divide this answer by 3, (following the dividing a decimal by a whole number method).

$$3\overline{)0.23659} = 3\overset{0.07886\dot{3}}{\overline{)0.236590}}$$

Notice the extra zero at the end of the dividend. This helps with calculations and makes no difference to the result.

Therefore $23.659 \div 300 = 0.0788633333333333\ldots$

As you can see, the quotient or answer has a three at the end that keeps on going. You only need to write one 3 with a dot over it ($\dot{3}$) to show that it repeats itself. This is an example of a **recurring decimal**.

Sometimes you need to change the 'look' of a decimal so you can complete your calculations.

For example

$5.6 \div 1000$ is the same as $0005.6 \div 1000$ (the extra zeros before the 5 are allowed in this case as it does not change the question but it helps us move the decimal point to the left).

Now divide by 1000 by moving the decimal point to the left three spaces.

Remember, the number of zeros in the divisor is the same as the number of spaces you move the decimal point to the left.

So $5.6 \div 1000$ becomes 0.0056.

Now you try these questions.

Calculate the following and indicate any recurring decimals appropriately:

a) $42.7 \div 100$

b) $64.91 \div 10000$

c) $4.63 \div 1000$

d) $230.5 \div 300$

e) $327.2 \div 700$

f) $225.62 \div 200$

g) $9.02 \div 120$

h) $16.32 \div 110$

i) $83.45 \div 1200$

j) $5.49 \div 20$

Chapter 7

Decimals and Fractions

There may be some math questions where you are required to convert decimals to fractions. At the front of this book was a diagram that looked like this:

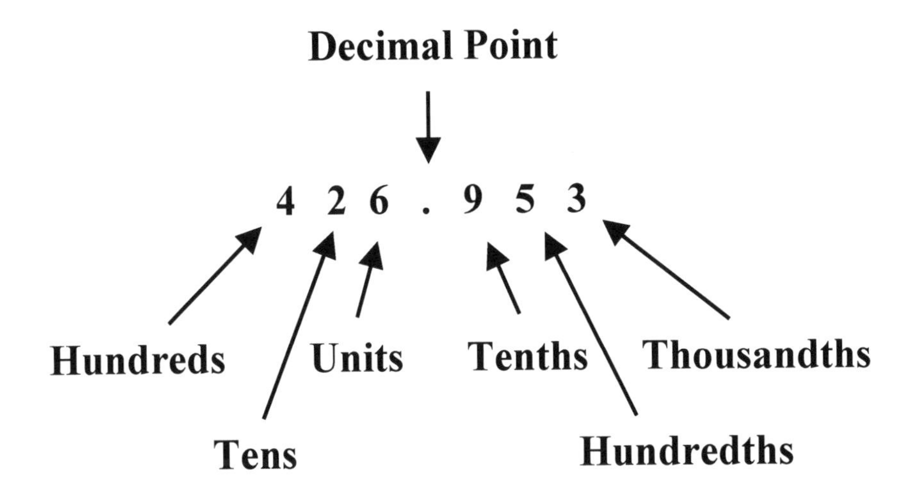

On the left of the decimal point there are whole numbers. These do not change. But on the right of the decimal point you will notice the values are tenths, hundredths, thousandths and this can keep going to ten thousandths, hundred thousandths, millionths, etc.

All these can be written as fractions.

The first digit after the decimal point is tenths.

So $0.1 = \frac{1}{10}$, $0.2 = \frac{2}{10}$, $0.3 = \frac{3}{10}$, and so on.

The second digit after the decimal point is hundredths.

So $0.01 = \frac{1}{100}$, $0.02 = \frac{2}{100}$, $0.03 = \frac{3}{100}$, $0.15 = \frac{15}{100}$,
$0.29 = \frac{29}{100}$, $0.84 = \frac{84}{100}$

The third digit after the decimal point is thousandths.

So $0.001 = \frac{1}{1000}$, $0.002 = \frac{2}{1000}$, $0.235 = \frac{235}{1000}$,
$0.999 = \frac{999}{1000}$, etc

Therefore you can see that the number of places to the right of the decimal point decides the number of zeros after the one in the denominator when you convert to a fraction.

The numbers on the left of the decimal point are the whole numbers, which makes the fraction a mixed number.

The decimal number written above, 426.953, can be written as the fraction $426\frac{953}{1000}$

The decimal number 2.36 is the same as the fraction $2\frac{36}{100}$, and this simplifies down to $2\frac{9}{25}$.

Don't forget to simplify fractions.

Now you try some.

Convert the following decimals to fractions, simplifying where necessary.

a) 6.1

b) 0.29

c) 0.00921

d) 1.0075

e) 7.3151

f) 8.224

g) 0.125

h) 1.000075

i) 4.985

j) 0.2354

There are times when you need to convert a fraction to a decimal in order to complete certain calculations.

Remember in the Easy Steps Math Fractions book it was said that a fraction is a division. Well, to convert $\frac{1}{5}$ to a decimal, just change it to $1 \div 5$ and put it in your calculator and it will give you the decimal answer 0.2. If you don't have a calculator, or if you're not allowed to use one, you need to do the conversion using either short division or long division depending on the numbers.

To convert $\frac{1}{5}$ to a decimal without a calculator you would do the following:

Step 1. You know that 5 doesn't go into 1, so you must put a decimal point and at least one zero after the 1 (see below).

$$5\overline{)1.0}$$

Step 2. Now start your division. Ask yourself "how many times does 5 go into 1"? The answer is zero so you write a zero on the line directly over the 1 and a decimal point directly over the decimal point below (remember that decimal points **must always** line up).

$$\begin{array}{r} 0. \\ 5\overline{)1.0} \end{array}$$

Step 3. Carry the 1 to the bottom zero to change it to a 10.

$$\begin{array}{r} 0. \\ 5\overline{)1.{}^{1}0} \end{array}$$

Step 4. Now ask yourself "how many times does 5 go into 10"? The answer is 2. Write the 2 over the line next to the decimal point (above the 0).

$$5\overline{)1.{}^{1}0}\quad \text{with quotient } 0.2$$

The answer is 0.2

Now you try these:

Convert the following fractions to decimals using your calculator.

a) $\frac{1}{4} =$

b) $\frac{2}{5} =$

c) $\frac{3}{8} =$

d) $\frac{37}{100} =$

e) $\frac{12}{25} =$

f) $\frac{43}{50}=$

Convert these fractions to decimals without your calculator.

g) $\frac{1}{2}=$

h) $\frac{3}{10}=$

i) $\frac{5}{8}=$

j) $\frac{7}{35}=$

k) $\frac{9}{20}=$

l) $\frac{3}{24}=$

You can check your answers with a calculator when you finish, not before.

Multiplication Tables

To make calculations really easy, learn your multiplications tables. Here is a set of multiplication tables from 1 x 1 to 12 x 12 to help you if you need it.

1 x 1 = 1	2 x 1 = 2	3 x 1 = 3	4 x 1 = 4
1 x 2 = 2	2 x 2 = 4	3 x 2 = 6	4 x 2 = 8
1 x 3 = 3	2 x 3 = 6	3 x 3 = 9	4 x 3 = 12
1 x 4 = 4	2 x 4 = 8	3 x 4 = 12	4 x 4 = 16
1 x 5 = 5	2 x 5 = 10	3 x 5 = 15	4 x 5 = 20
1 x 6 = 6	2 x 6 = 12	3 x 6 = 18	4 x 6 = 24
1 x 7 = 7	2 x 7 = 14	3 x 7 = 21	4 x 7 = 28
1 x 8 = 8	2 x 8 = 16	3 x 8 = 24	4 x 8 = 32
1 x 9 = 9	2 x 9 = 18	3 x 9 = 27	4 x 9 = 36
1 x 10 = 10	2 x 10 = 20	3 x 10 = 30	4 x 10 = 40
1 x 11 = 11	2 x 11 = 22	3 x 11 = 33	4 x 11 = 44
1 x 12 = 12	2 x 12 = 24	3 x 12 = 36	4 x 12 = 48

5 x 1 = 5	6 x 1 = 6	7 x 1 = 7	8 x 1 = 8
5 x 2 = 10	6 x 2 = 12	7 x 2 = 14	8 x 2 = 16
5 x 3 = 15	6 x 3 = 18	7 x 3 = 21	8 x 3 = 24
5 x 4 = 20	6 x 4 = 24	7 x 4 = 28	8 x 4 = 32
5 x 5 = 25	6 x 5 = 30	7 x 5 = 35	8 x 5 = 40
5 x 6 = 30	6 x 6 = 36	7 x 6 = 42	8 x 6 = 48
5 x 7 = 35	6 x 7 = 42	7 x 7 = 49	8 x 7 = 56
5 x 8 = 40	6 x 8 = 48	7 x 8 = 56	8 x 8 = 64
5 x 9 = 45	6 x 9 = 54	7 x 9 = 63	8 x 9 = 72
5 x 10 = 50	6 x 10 = 60	7 x 10 = 70	8 x 10 = 80
5 x 11 = 55	6 x 11 = 66	7 x 11 = 77	8 x 11 = 88
5 x 12 = 60	6 x 12 = 72	7 x 12 = 84	8 x 12 = 96

9 x 1 = 9	10 x 1 = 10	11 x 1 = 11	12 x 1 = 12
9 x 2 = 18	10 x 2 = 20	11 x 2 = 22	12 x 2 = 24
9 x 3 = 27	10 x 3 = 30	11 x 3 = 33	12 x 3 = 36
9 x 4 = 35	10 x 4 = 40	11 x 4 = 44	12 x 4 = 48
9 x 5 = 45	10 x 5 = 50	11 x 5 = 55	12 x 5 = 60
9 x 6 = 54	10 x 6 = 60	11 x 6 = 66	12 x 6 = 72
9 x 7 = 63	10 x 7 = 70	11 x 7 = 77	12 x 7 = 84
9 x 8 = 72	10 x 8 = 80	11 x 8 = 88	12 x 8 = 96
9 x 9 = 81	10 x 9 = 90	11 x 9 = 99	12 x 9 = 108
9 x 10 = 90	10 x 10 = 100	11 x 10 =110	12 x 10 = 120
9 x 11 = 99	10 x 11 = 110	11 x 11 = 121	12 x 11 = 132
9 x 12 = 108	10 x 12 = 120	11 x 12 = 132	12 x 12 = 144

Answers

Place Values

a) $\frac{5}{100}$ b) $\frac{3}{10}$ c) 20 d) $\frac{6}{1000}$ e) $\frac{8}{10000}$ f) 0.3 g) 0.07
h) 0.15 i) 2.23 j) 15.145

Comparing Decimals

a) 4.56, 4.65, 5.46, 5.64, 6.45, 6.54

b) 3.40, 3.41, 3.5, 3.7, 3.73, 3.75

c) 4.6, 4.67, 4.677, 4.706, 4.71

d) 3.116, 3.616, 3.66, 3.661

e) 1.874, 1.876, 2.37, 2.45, 3.098, 3.102, 4.5, 4.983

f) the incorrect answer is **B** 6.646 > 6.664

g) the correct answer is **C** 0.7194 > 0.7149

Rounding Decimals

a) 18.6 b) 7.6 c) 0.6 d) 9.4 e) 7.9 f) 8.0 g) 5.79 h) 12.57 i) 32.46 j) 400.26 k) 27 l) 5 m) 84 n) 121 o) 76 p) $27.00 q) $15.00 r) $424.00 s) $21.00

Addition and Subtractions of Decimals

a) 20.48 b) 25 c) 84.11 d) 8.32 e) 68.236 f) 261.05
g) 94.03 h) 190.932 i) 16.774 j) 20.963

Multiplying Decimals with Whole Numbers

a) 14.0 b) 110.7 c) 81.6 d) 184.65 e) 378.58 f) 7.858
g) 63.1845 h) 368.6 i) 1295.36 j) 104.70

Multiplying a Decimal with Another Decimal

a) 0.32 b) 0.747 c) 9.72 d) 2.0812 e) 1.5072 f) 32.045
g) 0.021 h) 0.0002 i) 1999.33 j) 13.46

Multiplying a Decimal with a Multiple of 10

a) 836 b) 35 c) 4630 d) 691.5 e) 22890 f) 676.86
g) 4680 h) 8000 i) 120 j) 75600

Dividing Decimals with Whole Numbers

a) 0.586 b) 5.29 c) 0.706 d) 2.063 e) 0.30 f) 0.224
g) 1.7 h) 0.85 i) 4.125 j) 0.22

Dividing Decimals with Another Decimal

a) 122.1 b) 8.5 c) 8.3 d) 5610 e) 8700 f) 0.467
g) 214.167 h) 57.290 i) 926.4 j) 0.97

Dividing a Decimal with a Multiple of 10

a) 0.427 b) 0.006491 c) 0.00463 d) 0.768(3)
e) 0.467(428571) f) 1.1281 g) 0.0751(6) h) 0.148(36)
i) 0.069541(6) j) 0.2745

Converting a Decimal to a Fraction

a) $6\frac{1}{10}$ b) $\frac{29}{100}$ c) $\frac{921}{100000}$ d) $1\frac{3}{400}$ e) $7\frac{3151}{10000}$ f) $8\frac{28}{125}$
g) $\frac{1}{8}$ h) $1\frac{3}{40000}$ i) $4\frac{197}{200}$ j) $\frac{1177}{5000}$

Converting a Fraction to a Decimal

a) 0.25 b) 0.4 c) 0.375 d) 0.37 e) 0.48 f) 0.86 g) 0.5
h) 0.3 i) 0.625 j) 0.2 k) 0.45 l) 0.125

Glossary of Useful Terms

Sum refers to addition. The sum of two numbers is the answer of one number **plus** another number. E.g. the sum of 2 and 6 is 8, ($2 + 6 = 8$).

Difference refers to subtraction. The difference between two numbers is the answer of one number **minus** another number. E.g. the difference between 6 and 2 is 4, ($6 - 2 = 4$).

Product refers to multiplication. The product of two numbers is the answer of one number **times** another number. E.g. the product of 2 and 6 is 12, ($2 \times 6 = 12$).

Quotient refers to division. The quotient is the answer of one number being **divided** by another number. E.g. the quotient of 6 and 2 is 3 ($6 \div 2 = 3$).

An **integer** is a positive or negative whole number, including zero, which does not have any fractional components.

A **digit** is a single numeral from 0 to 9. Putting digits together makes larger numbers e.g. 325.

A **recurring** decimal or **repeating** decimal is one with a repeating pattern. The fraction $\frac{1}{3}$, when written as a decimal is 0.3333333333..., the threes go on forever. To simplify things a recurring decimal is written with a dot over the recurring digit i.e. $0.\dot{3}$. In some countries the method is to put a bar over the recurring decimal i.e. $0.\overline{3}$, and in other countries the recurring decimal will have parentheses around it i.e. $0.(3)$.

Some other examples of recurring or repeating decimals are:

$$\frac{9}{11} = 0.\dot{8}\dot{1} \; or \; 0.\overline{81} \; or \; 0.(81)$$ for 2 recurring decimals

and

$$\frac{1}{41} = 0.\dot{0}243\dot{9} \; or \; 0.\overline{02439} \; or \; 0.(02439)$$

for 5 recurring decimals.

Notice how the dots are only placed over the first and last recurring digits and not over all digits.

Thank you for reading!

Dear Reader,

I hope you found this **Easy Steps Math – Decimals** book useful, either for yourself or for your children.

The **Easy Steps Math** series began as a set of math notes that I used in class for my students to copy from the board. I found that doing this helped the students in at least two ways. Firstly, with their homework, because they didn't forget how to do the work that was explained in class, and secondly, with their results, because they used the notes to study for tests and exams.

Where students in other classes were getting detentions for non-completion of homework, my students were getting homework done, their results were improving and they were enjoying math.

Students from other classes, even older students, were thanking me for my notes, as they were copying them from their peers because they found them so easy to follow and learn from. During a parent/teacher conference, one parent also thanked me because of how his child was able to easily learn the work, and that he, as a teacher, was using my notes in his classes, in his school.

This is when I realised that these notes would benefit many more students if they were published. Thus we are at this point.

I welcome any comments you have about this **Decimals** book. Tell me what you liked, loved, or even hated about it. I'd be happy to hear from you. You can email me at robwatchman@gmail.com.

Finally, I would like to ask a favour. I would appreciate it if you would write a review of this book so that others can get an idea of how helpful it may be for them or their children. You would be aware that reviews are hard to come by because many readers don't go back to where they purchased their books.

So if you have the time, here is the link to my author's page on Amazon. You can find all of my other books here also: http://amzn.to/1rlW6gr.

Thank you so much for reading the **Easy Steps Math – Decimals** book and for spending your time with me.

In Gratitude,

Robert Watchman

Made in the USA
Middletown, DE
27 May 2020

96076873R00033